IMPROVISED EXPLOSIVES For Use In DETONATORS

DESERT PUBLICATIONS

IMPROVISED EXPLOSIVES
FOR USE IN DETONATORS

ISBN: 0-87947-220-0

DESERT PUBLICATIONS
Cornville, Arizona 86325

Table of Contents

Summary

A program has been completed which was directed toward the development of field expedient methods of preparing primary explosives for use in indigenous detonators.

Refined procedures were developed for three primary explosives:

(1) Diazodinitrophenol
(2) Lead picrate
(3) Tetramminecopper (II) chlorate (TAC)

In addition, a method possibly conducive to field use of lead azide was developed.

Other accomplishments included the development of methods for preparing the following secondary explosives, intermediates, and aids:

(1) Picric acid
(2) Reclamation of RDX from C4
(3) Nitrourea
(4) Sodium (or potassium) nitrite
(5) Lead monixide
(6) Copper sulfate
(7) Hydrazine sulfate
(8) Potassium chlorate string fuze

It should also be mentioned that a number of additional explosives were investigated, many of them warranting future consideration.

The development of indigenous detonators, while not constituting the major emphasis of the program, nevertheless required that a survey be performed in order to arrive at suitable hardware for testing the candidate primaries. Cartridge brass and relatively heavy wall copper tubing were found satisfactory as detonator case materials.

I. Introduction

General

A 12 man-month program has been performed with the objective of developing field expedient methods for preparing primary explosives for use in indigenous detonators.

Technical Approach

The approach to the task of developing field methods for preparing primary explosives was guided by three considerations:

(1) The existence of field methods for preparing four known compounds, the fulminates of mercury and silver, silver acetylide/silver nitrate (double salt), and hexamethylene triperoxidediamene (HMTD).

(2) The existence of a significant number of additional well-known primary explosives which could lend themselves to field expedient processes.

(3) The availability of much information concerning rather obscure compounds potentially capable of functioning as primary explosives.

In addition to these considerations the general program goal was the development of expedient procedures for the purpose of fabricating selected detonator systems using indigenous materials.

Program Outline

The program was divided into four operation-oriented tasks:

(1) Fabrication of indigenous detonators
(2) Preparation of primary explosives
(3) Preparation of booster explosives
(4) Testing of detonators

The first three items were dependent on availability of chemicals, utensils, and processes which were suitable for use by novice personnel. Task (4) was a testing criteria operation associated with the evaluation of each candidate material under tasks (1), (2), and (3).

II. Detonator Design

Functional evaluation of the explosives prepared as described in the following sections of this document required the design and fabrication of suitable detonators from indigenous material. Consequently, part of the initial effort was devoted to the development of appropriate detonator hardware using the experimental explosives. The development was conducted on an empirical basis during approximately the first half of the program and consisted of (1) materials selection, (2) fabrication, and (3) testing.

Materials Selection

Materials evaluated for detonator construction encompassed a broad variety of metal entities.

a. Case Material - Initial testing was performed using No. 6 blasting cap cases, with subsequent studies employing various diameters and types of metal tubing. In the latter stages of the program it was determined that cartridge brass could be utilized quite successfully.
b. Fuse Material - The fusing employed throughout the program was a firecracker safety fuse; however, it was determined that cotton stringsaturated with potassium chlorate should afford an acceptable item.

Detonator testing consisted of initiating C-4 or KAL and employing metal plates for witness of detonator success.

Fabrication Methods

The testing using No. 6 blasting cap cases indicated more confinement was necessary; therefore, copper tubing of more substantial wall thickness was evaluated. It was determined at this stage that initiation of C-4 would be difficult when using only a primary explosive.

Subsequently, compound detonators were fabricated and evaluated with success. Initially, these were prepared by inserting a thick-walled copper tube (0.250-inch O.D. x 0.47-inch wall) filled with primary explosive into a No. 6 cap case containing booster explosive (RDX or picric acid). This concept was later simplified by merely employing a single metal tube for both explosives. A further successful modification consisted of using various sizes of cartridge brass with 30-06 cases being the usual hardware. It was determined that from an overall standpoint the 30-06 cartridge brass or the 0.25 copper tubing (0.030-inch O. D. and 0.047-inch wall thickness) were the optimum materials for indigenous detonators. Cartridge brass offers the advantage of not requiring crimping. The various types of metal employed for detonator housing are given in Table I.

The loading of explosives into the housing was initially performed using a car jack and appropriate timber support; however, it was later found that hand pressing with a wooden rod was effective and less cumbersome. With regard to tubing it was found advantageous to crimp the fuse in place prior to loading of the explosive. Loading of the cartridge brass cases was accomplished by placing fuse in the flash hole, which had been increased in size with a nail, then hand loading the primary, booster, and C-4* charges in that order toward the fuse end.

A detailed procedure for the field expedient fabrication of indigenous detonators is provided in Appendix A of this report.

Fuse Study

Since it was questionable whether or not an expedient process existed for preparing suitable linear fusing, a brief study was undertaken near the termination of the program.

---

* The C-4 was incorporated into the cartridges (30-06) in order to fill the remaining void. Where KAL was used, the void was filled with paper wadding and the detonator immersed in the receptor.

TABLE I
Hardware Employed For Detonator Cases

Metal	Outside Diameter (inch)	Wall Thickness (inch)
Copper	0.272*	0.011*
Copper	0.250	0.047
Copper	0.250	0.030
Copper	0.375	0.032
Copper	0.500	0.035
Brass	0.344	0.015
Aluminum	0.375	0.022
Cartridge, Brass	.30-06 case	
Cartridge, Brass	.308 case	
Cartridge, Brass	.222 case	

*No. 6 blasting cap cases, closed end.

Four types of cotton string fuse were studied:

(1) Potassium Nitrate
(2) Potassium Chlorate
(3) Lead Picrate
(4) TAC

a. The potassium nitrate fuses were prepared from saturated water and methanol solutions and cotton string. Both items burned very slowly with no flame.
b. Potassium chlorate fuses (using both water and methanol solutions) exhibited burning characteristics superior to the other candidates. The burning rates of ten 6-inch strands varied from 0.95 to 2.5 seconds/ inch with the average being 1.57.
c. Lead picrate and TAC fuses were prepared from liquid slurries of the explosives and exhibited somewhat erratic burning characteristics.

III. Primary Explosives Investigation

Candidate Materials

A rather large number of candidates were investigated (see Table II), the selection varying from well known primary explosives to rather obscure chemical curiosities.

Group I (Table II) represents established initiating explosives, and Group II consists of primary explosives having less history than those in Group I. However, Group III explosives were new entries in the primary respect and were studied in an attempt to provide novel solutions to the program requirements. While the explosives developed for field application under this program were for the most part representatives of Group I (DDNP and lead picrate) and in one instance Group II (tetramminecopper (II) chlorate), Groups II and III are considered excellent sources of applicable explosives.

Lead Picrate ($C_6H_2(NO_2)_3OPb$)

The metal picrates warranted investigation due to their known primary explosives characteristics and relative ease of preparation. Lead picrate was chosen because of its considerable history and the fact that it can be prepared by a number of simple methods.

A method has been developed by which picric acid can be prepared from aspirin, sulfuric acid, and potassium nitrate in less than two hours. Essentially, the procedure consists of extracting pure acetylsalicylic acid from the tablet form with alcohol and subsequently evaporating the alcohol. The dry residue is dissolved and heated in concentrated sulfuric acid for a short period, and then nitrated by the portion-wise addition of sodium or potassium nitrate. After a brief reaction period the reaction mixture is added to cold water and the picric acid isolated by filtration. Yields of pure product are

TABLE II
Candidate Materials Investigated For Primary Explosive Properties

	Impact (Notes) Sensitivity (cm)
GROUP I - ESTABLISHED PRIMARY EXPLOSIVES	
Diazodinitrophenol (DDNP)	16
Lead Picrate	15
Lead Azide	
Mannitol hexanitrate	
Nitrogen tetrasulfide	
GROUP II - UNPROVEN PRIMARY EXPLOSIVES	
Tetramminecopper (II) chlorate (TAC)	10
Triacetonetriperoxide	2-3
Fulminating silver (and gold)	
Polyol complexes of lead chlorates	
Chloratotrimercuriacetaldehyde	5
Mercarbid and salts	
GROUP III - NEW PRIMARY EXPLOSIVES	
Lead nitranilate	
Trishydrazinezinc (II) nitrate	
Diamminesilver (I) chlorate	
Complexes of semicarbazide and biurea (Hydrazodicarboxamide)	
Salts of nitrourea	

NOTES: (1) Values obtained on compounds prepared by refined field methods.

(2) Sensitivity values refer to 10 per cent fire level obtained obtained using the Standard Bureau of Mines test apparatus with a 2.0 Kg mass.

about ninety percent of theoretical and the total time required is about two hours.

The lead salt was prepared by two methods:

(1) Reaction of an ammoniacal solution of picric acid with a lead salt, and

(2) Blending a methanolic slurry of picric acid and lead monoxide (litharge), the latter having been prepared from potassium nitrate and lead.

Method (2) was ultimately chosen for reasons of expediency.

When lead picrate was used with a booster, such as RDX (reclaimed from C-4) or picric acid, C-4 was successfully detonated. However, attempts to initiate C-4 without a booster were unsuccessful. As indicated in Table III relatively large quantities of lead picrate (2-3 g) were required for fabrication of successful detonators. Cartridge brass or copper tubing (0.25-inch O.D.) were successfully employed in detonators, with RDX being the better choice of booster.

While lead picrate is potentially an excellent initiator, the compound is of an explosive type that required confinement, or a minimum critical column length, or both, depending on the extent of either parameter. Nevertheless, the material is easily prepared in quantity in a short time and is considered relatively safe compared to many primary explosives.

Diazodinitrophenol (DDNP)

A cursory examination of previous methods of preparing DDNP might have resulted in the prediction that it would not lend itself to field manufacture. On the contrary, however, a field method for reducing picric acid to picramic acid and a simple method for preparing sodium nitrite (required to convert picramic acid to DDNP) has provided an expedient process to this primary.

The reduction process entails the reaction of a hot aqueous sulfur/sodium hydroxide (lye) solution with picric acid. Picramic acid is isolated from impurities by filtration and acidification steps. Sodium nitrite is prepared by fusion of lead

TABLE III
Summary of Detonator Testing
Lead Picrate

Weight Lead Picrate (grams)	Booster Weight (grams)	Detonator Case Size (see Table I) (inch O.D.)	Results
Detonator Case: Brass, 30-06 Cartridge Case, Primary on Cap End.			
3.0	1.0 RDX	-	Initiated C-4
3.0	1.0 RDX	-	Initiated C-4
3.0	1.5 RDX	-	Initiated C-4
2.5	1.5 PA (Note)	-	Initiated C-4
3.0	1.0 PA	-	Initiated C-4
2.0	1.5 PA	-	Initiated KAL
2.0	1.0 RDX	-	Initiated KAL
2.0	----	-	Low order detonation of KAL*
Detonator Case: Brass 30-06 Cartridge Case, Booster on Cap End.			
3.0	1.0 PA	-	PA consumed, C-4 not initiated
3.0	1.0 RDX	-	RDX consumed, C-4 not initiated
3.0	1.0 RDX	-	RDX consumed, C-4 not initiated
3.0	1.0 RDX	-	RDX consumed, C-4 not initiated
3.0	2.0 RDX	-	RDX consumed, C-4 not initiated
3.0	1.0 C-4	-	RDX consumed, C-4 not initiated
3.0	----	-	
Detonator Case: Copper Tubing.			
2.5	1.5 RDX	0.375	Initiated C-4
2.5	1.0 RDX	0.375	Initiated C-4
2.5	1.0 RDX	0.25	Initiated C-4
3.0; 4 trials	1.0 RDX	0.25	Initiated C-4
3.0	1.0 RDX	0.25	Initiated C-4
3.0	1.0 RDX	0.25	Initiated C-4
3.0	1.0 RDX	0.25	Initiated C-4
Detonator Case: Aluminum Tubing			
3.0	2.0 RDX	0.375	Initiated C-4
2.1	1.5 PA	0.375	Did not initiate PA or C-4
2.3	1.5 RDX	0.375	Initiated C-4
2.5	----	0.25	Did not initiate C-4

Note: PA - Picric acid

Table III
(Continued)

Weight Lead Picrate (grams)	Booster Weight (grams)	Detonator Case Size (see Table I) (inch O.D.)	Results
		Detonator Case: Brass Tubing	
2.5	0.5 tetryl	0.344	Did not initiate tetryl or C-4
1.0	1.0 manni – to hexanitrate	0.344	Did not initiate MH or C-4
2.5	0.5 manni- to hexanitrate	0.344	Did not initiate MH or C-4
0.8 PbPic 0.2 $NaClO_3$	0.5 RDX	0.344	Did not initiate RDX or C-4
1.0	1.0 RDX	0.344	Did not initiate RDX or C-4
1.5	1.0 RDX	0.344	Did not initiate RDX or C-4
2.5	0.5 RDX	0.344	Did not initiate RDX or C-4
0.5	0.5 PA	0.344	Did not initiate PA or C-4
0.5	1.0 PA	0.344	Did not initiate PA or C-4
0.7	0.5 PA	0.344	Did not initiate PA or C-4
1.0	1.0 PA	0.344	Did not initiate PA or C-4
2.5	0.3 PA	0.344	Did not initiate PA or C-4
0.5	----	0.344	Did not initiate C-4
1.0	----	0.344	Did not initiate C-4
1.5	----	0.344	Did not initiate C-4
2.0	----	0.344	Did not initiate C-4
3.0	----	0.344	Did not initiate C-4
		Detonator Case: No. 6 Blasting Cap Case	
1.0 + 1.3**	----	0.25 copper	Did not initiate C-4
1.0 + 0.5**	----	0.25 copper	Did not initiate C-4
0.4	0.5 PA	-	PA consumed, did not initiate C-4
0.4	0.5 PA	-	PA consumed, did not initiate C-4
0.2	----	-	Did not initiate C-4
0.4	----	-	Did not initiate C-4
0.9	----	-	Did not initiate C-4

*Indicated by the presence of residual aluminum after the test.
**First number denotes grams in copper tube; second number in blasting cap.

metal with sodium or potassium nitrate and subsequent leaching. It is important to note that a by-product of this reaction is lead monoxide, which is required to prepare a lead picrate. The overall reaction scheme is presented below.

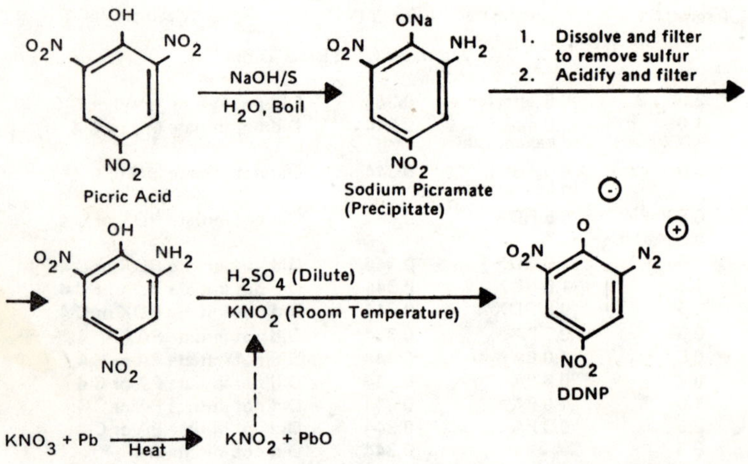

The initiation capabilities of DDNP are superior to those of lead picrate in that quantities as low as 0.4 to 0.5 grams have been successfully employed in detonating C-4, using either RDX or picric acid boosters. (cf., Table IV). Both cartridge brass and copper tubing were used with success in the detonator fabrications. The important factors to be emphasized with regard to DDNP are:

(1) Dryness
(2) Confinement and compaction

It was found that drying for one hour at 80° C was not sufficient, but that two hours at the same temperature afforded a suitable product. While confinement is necessary, it appears to be not as critical for lead picrate. This conclusion is evident from the successful tests employing No. 6 blasting cap cases (cf. Table IV). As noted in Table III, lead picrate did not function successfully with this hardware.

Tetramminecopper (II) Chlorate (TAC)

The chemical class of coordination compounds offers unlimited possibilities in the area of explosives chemistry. TAC was chosen for refinement since the literature mentions it to be capable of initiating TNT. [1] This explosive, from at least one viewpoint, is considered the best candidate developed during the program. In the initial studies, isolation of the pure compound presented the major difficulty; however, this was overcome by appropriate choice of solvents and elimination of the presence of significant quantities of water.

Essentially, the laboratory procedure consists of dissolving sodium chlorate and copper sulfate in methyl or ethyl alcohol, heating the solution on a water bath, and removing by filtration the sodium sulfate formed. The filtrate is treated with gaseous ammonia which precipitates the product as brilliant blue needles.

The ammonia gas was prepared by heating a tubulated bottle containing household ammonia in a hot water bath. The overall reaction can be represented as follows:

$$CuSO_4 + NaClO_3 \xrightarrow[\text{heat}]{CH_3OH} \underset{\text{in solution}}{Cu(ClO_3)_2} \xrightarrow[]{\substack{NH_3 \\ \text{(Gas)}}} \underset{\text{precipitate}}{[Cu(NH_3)_4]\,(ClO_3)_2}$$

$$+$$

$$\underset{\substack{\text{precipitate} \\ \text{(removed by filtration)}}}{Na_2SO_4}$$

From an explosive aspect, TAC appeared superior to the previously discussed organic based explosives in that hand

---

(1) Federoff, Basil T. and Sheffield, Oliver E., Encyclopedia of Explosives and Related Items, p. A283, Volume I, Picatinny Arsenal, 1960.

TABLE IV
Summary of Detonator Testing
Diazodinitrophenol

Weight Primary (grams)	Booster Weight (grams)	Detonator Case Size (see Table I) (inch O.D.)	Results
Detonator Case: Brass, 30-06 Cartridge, Primary on Cap End.			
1.0	1.0 RDX	30-06	Initiated C-4
0.4	1.5 PA	30-06	Initiated C-4
0.5	1.0 RDX	30-06	Initiated KAL
0.5	1.5 PA	30-06	Initiated KAL
1.0	----	30-06	Initiated KAL
Detonator Case: Copper Tubing.			
1.0	1.0 RDX	0.25	Initiated C-4, primary dried 2 hours (80° C)
0.7	1.0 RDX	0.25	Burned, improper crimping
0.7	1.0 RDX	0.25	Initiated C-4
0.7	1.0 RDX	0.25	Initiated C-4
0.6; 3 trials	1.0 RDX	0.25	Initiated C-4
0.5	1.5 RDX	0.25	Initiated C-4
0.5; 4 trials	1.0 RDX	0.25	Initiated C-4
0.74	1.0 PA	0.25	Initiated C-4, primary dried 2 hours (80° C)
0.7	1.1 PA	0.25	Initiated C-4, primary dried 2 hours (80° C)
0.7; 2 trials	1.0 PA	0.25	Initiated C-4, primary dried 2 hours (80° C)
0.6	1.0 PA	0.25	Initiated C-4, primary dried 2 hours (80° C)
0.5; 2 trials	1.0 PA	0.25	Initiated C-4, primary dried 2 hours (80° C)
0.7	0.7 PA	0.25	Initiated C-4, primary dried 2 hours (80° C)
1.0; 2 trials	2.0 PA	0.25	Did not initiate C-4, primary dried 1 hour (80° C)
1.0	----	0.25	Did not initiate C-4, primary dried 1 hour (80° C)
0.25	1.0 RDX	0.485	Did not initiate RDX or C-4
1.1	----	0.50	Did not initiate C-4
1.5	1.5 RDX	0.25*	Initiated C-4, primary dried 2.5 hours (80° C)

TABLE IV
(Continued)

Weight Primary (grams)	Booster Weight (grams)	Detonator Case Size (see Table I) (inch O.D.)	Results
1.5	1.5 RDX	0.25*	Initiated C-4, primary dried 2.5 hours (80° C)
1.0	1.5 RDX	0.25*	Initiated C-4, primary dried 2.5 hours (80° C)
1.1	1.0 RDX	0.25*	Initiated C-4, primary dried 2.5 hours (80° C)
1.0	1.0 RDX	0.25*	Initiated C-4, primary dried 2.5 hours (80° C)
0.8	1.0 RDX	0.25*	Initiated C-4
1.0; 3 trials	1.5 RDX	0.25*	Did not initiate RDX or C-4 primary dried 1 hour (80° C)
1.0	2.0 PA	0.25*	Did not initiate PA or C-4, primary dried 1 hour (80° C)
1.0	2.0 PA	0.25*	Did not initiate PA or C-4, primary dried 1 hour (80° C)
1.8	----	0.25*	Did not initiate C-4
0.3	1.0 PA	0.25*	Did not initiate PA or C-4
1.0	----	0.25*	Did not initiate C-4
0.3	1.0 tetryl	0.25*	Did not initiate tetryl or C-4
0.3	1.0 RDX	0.25*	Did not initiate RDX or C-4
		Detonator Case: Aluminum Tubing	
1.0: 2 trials	----	0.375*	Did not initiate C-4

*Fired with a No. 6 blasting cap.

tamping of the material was not required to achieve initiation of RDX or picric acid boosters in the detonator testing (cf. Table V). Both cartridge brass and copper tubing were employed with success in the fabrication of detonators for C-4 and KAL. Attempts to initiate C-4 with TAC without a booster however, have not been successful.

Additional Primary Explosive Candidates

The compounds described previously in this section represent materials which are proven initiators and for which field preparation methods have been developed. This portion of the section presents a discussion of the remaining candidates. Their potential and the degrees of effort expended on them varies considerably. The more promising materials will be discussed initially.

Lead Azide

Since this compound is favored by far over other initiating explosives in ordnance applications, a study was made of the possibility of developing a field manufacturing procedure. At the present time a procedure has been developed which could possibly be carried out by the novice. However, it is not considered as expedient as those previously discussed in this section. The preparation is schematically presented on page 20.

Lead azide prepared in accordance with the above sequence proved capable of initiating RDX. However, the quality of the lead compound is not the major point of consideration at this juncture, but rather the method of preparation. The method is somewhat time consuming and does require techniques probably requiring some practice; however, a limited effort would simplify certain aspects of the manufacture.

Mannitol Hexanitrate

This explosive is prepared with relative ease by nitrating mannitol with mixed concentrated nitric and sulfuric acids at 0° C. However, the problem of mannitol procurement and

TABLE V
Summary of Detonator Testing
Tetrammine Copper (II) Chlorate (TAC)

Weight Primary (grams)	Booster Weight (grams)	Detonator Case Size (see Table I) (inch O.D.)	Results
Detonator Case: Brass, 30-06 Cartridge, Primary on Cap End			
1.0	1.0 PA	30-06	Initiated C-4
1.0	0.5 PA	30-06	Initiated C-4
1.0	0.5 RDX	.222	Initiated C-4
1.0; 6 trials	1.0 RDX	30-06	Initiated C-4
1.0	1.0 RDX	30-06	Initiated C-4, TAC loose not tamped
1.0	----	.308	Did not initiate C-4
1.0	----	30-06	Did not initiate C-4
1.0	----	30-06	Low order detonation of KAL
2.0	----	30-06	Low order detonation of KAL
Detonator Case: Tubing			
1.0	1.0 PA	.375 copper	Initiated C-4
1.0	1.0 PA	.375 copper	Initiated C-4
1.0; 2 trials	1.0 RDX	.375 copper	Initiated C-4
3.0	----	.375 copper	Did not initiate C-4

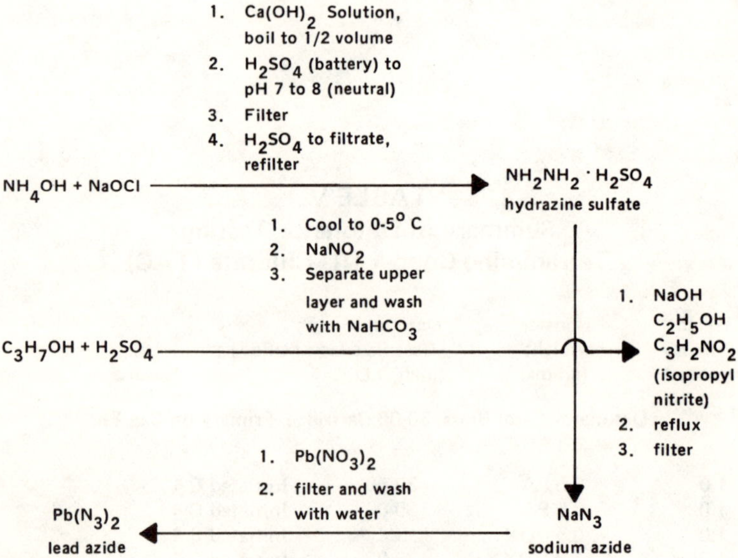

Lead Azide Preparation

relatively high degree of confinement required to achieve detonation have precluded its extensive investigation. Preliminary testing resulted in failure of hexanitromannitol to initiate RDX. However, the compound has been reported applicable as a primary explosive in detonators[1].

Triacetonetriperoxide (TATP)

In the latter stages of the program a brief study was made of the preparation of this primary explosive. It was determined that six percent hydrogen peroxide could be successfully used in the preparation rather than the thirty percent reported in the literature[2]. In addition to the peroxide, acetone and

---

(1) Burrows, L. A., U. S. Patent 2,427,899 (23 September 1947).

(2) Federoff, Basil T. and Sheffield, Oliver E., Encyclopedia of Explosives and Related Items, p. A42, Volume 1, Picatinny Arsenal, 1960.

concentrated sulfuric acid are required, with reaction conditions being at about 5° C.

The product is quite impact sensitive (2-3 cm, 2 Kg BM apparatus) and has the disadvantage of being quite volatile, dictating that it be used soon after preparation. Although time did not permit its full evaluation, the compound may reveal some advantages through limited additional study.

Lead Nitranilate

A rather unique series of reactions led to the preparation of the high compound sodium nitranilate by the following sequence of reactions:

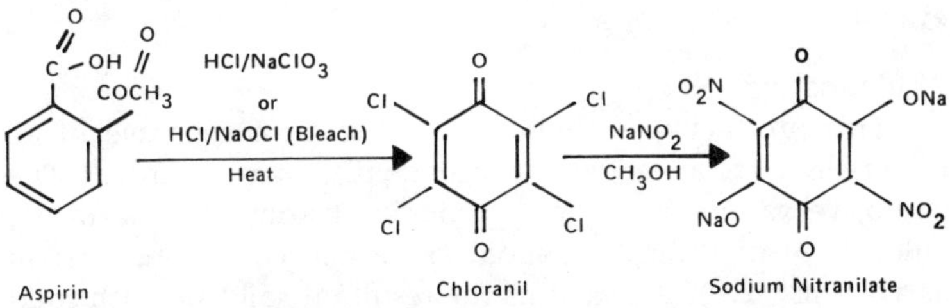

This sequence of reactions was easily performed as was the subsequent preparation of the lead complex employing lead nitrate. Although the lead compound exhibits "primary activity" on flame contact, attempts to initiate RDX in 0.25 copper tubing was unsuccessful. Further study might disclose proper methods for utilizing the explosive in detonators.

Salt of Nitrourea

The silver and mercury (II) salts of nitrourea exhibit greater flame activity than the parent explosive, but they do not burn in a manner as to lend them to primary explosives use. However, it was found the burning activity of nitrourea could be considerably enhanced by the addition of bleaching powder. The major interest in this study is the possible use of nitrourea as a secondary explosive. This possibility is discussed in detail in the section concerning booster explosives.

Chloratotrimercuriacetaldehyde

This compound [1] was easily prepared by passing acetylene into a solution of mercuric nitrate and sodium chlorate. The grey solid product detonated at about 150° C and fired at five cm on the 2 Kg Bureau of Mines Apparatus. Further study was terminated in view of the sensitivity. However, a binder investigation might result in a practical material.

Mercarbid and Salts of HOHg ($OHg_2C_2Hg_2O$) HgOH

The reaction of yellow mercuric oxide, potassium hydroxide and ethyl alcohol is reported to yield the parent compound[2]. Attempts to prepare the nitrate gave inconclusive results; however, the investigation was made near the termination of the program. Consequently, a more complete evaluation was not possible.

"Fulminating" Silver

This product was found to be the most sensitive of all materials investigated and consequently was discarded. The explosive was prepared several times by dissolving silver nitrate chloride or oxide in household ammonia (containing dextrin as a desensitizer), and treating the resultant solution with solid potassium hydroxide. In all cases, black products were obtained which were exploded on handling in the moist state.

Nitrogen Sulfide (N_4S_4)

Efforts to prepare this compound were curtailed due to unsuccessful attempts to prepare the required precursor, sulfur dichloride. It is of interest that the attempted reaction of sulfur and calcium hypochlorite (bleaching powder) resulted in ignition on mixing, thus affording an indigenous contact ignition system.

---

(1) Federoff, Basil T. and Sheffield, Oliver E., Encyclopedia of Explosives and Related Items, p. c 209, Volume 2, Picatinny Arsenal, 1962.

(2) Beilstein, Band II, p. 562, (1960).

Various Coordination Compounds

The reactions of a number of ligands with various metal nitrates and chlorates were investigated with the intention of developing new primary explosive. One of these, tetrammine-copper (II) chlorate (TAC), had resulted in a successful detonator as described prevoiously. Other candidates were studied to lesser degrees and with additional effort, several compounds could attain prominence equal to TAC.

a. Glycerol Complex of Lead Chlorate. Probably the most feasible of these compounds would be the glycerol complex of normal and basic lead chlorates, a material proposed in the past for use in initiators. Attempts to prepare the required lead chlorates in a field manner did not meet with satisfactory results, a situation undoubtedly rectifiable by additional work.

b. Salts of Hydrazine. The development of a field method for preparing salts of hydrazine for use in preparing lead azide, prompted the brief investigation of trishydrazine-zinc (II) nitrate. The compound was easily prepared from hydrazine and zinc nitrate (made from zinc and nitric acid). However, it did not exhibit activity characteristic of a primary explosive. Hydrazine was also reacted with urea (fertilizer) to give the ligands hydrazine dicarbox-amide and semicarbazide. Several preliminary experiments employing metal nitrates and these ligands have resulted in inconclusive results; however, the semicarbazide, in particular, does offer future potential for synthesis of primary explosives and as a propellant additive.

c. Diamminesilver (I) Chlorate. Ammonia and silver nitrate were reacted to yield an impure form of diamminesilver (I) chlorate, which exhibited only mild deflagrating characteristics and was subsequently deleted from further study. It should be mentioned that owing to its relative insolubility, silver chlorate is easily prepared from aqueous silver nitrate and sodium chlorate solutions and possibly would offer future possibilities.

IV. Booster Explosives Investigation

This class of compounds was not the prime investigational goal of the program. However, their existence as precursors for primary explosives and the requirement of boosters has included them in the program.

Picric Acid

As previously described, this compound is a requisite intermediate in the preparation of lead picrate and DDNP. A facile synthesis of this material was developed employing sulfuric acid, potassium nitrate and aspirin. This material not only sufficed for preparing the primary explosives but was satisfactory as a booster when properly dried (80-100° C) for two hours.

RDX

A simple process employing automotive gasoline permits the near quantitative extraction of this explosive from C-4. The reclamation process was performed by extracting the oils from C-4 with gasoline at ambient conditions and collecting the residual RDX in a filter. After drying, the product is an effective booster explosive for detonators.

Nitrourea

While no explosive evaluation was performed on this material, its ease of preparation and stable nature warrant consideration. The compound is easily prepared by the addition of urea nitrate to cold concentrated sulfuric acid and subsequent addition to ice to precipitate the product. Urea nitrate is prepared from urine by a number of simple, well established procedures using nitric acid.

The compound as it stands is a high explosive or it can be blended with aluminum powder for increased output. Preliminary experiments under this program indicated its activity can be enhanced by the addition of bleaching powder. While beyond the scope of the detonator program, the compound should definitely receive future consideration for secondary explosive applications.

V. Conclusions & Recommendations

The results of the development program demonstrated that field expedient detonators can be prepared using the following indigenous materials:

(1) Detonator Case - Either 30-06 cartridge or heavy-walled copper tubing (0.25-0.375 O. D.).
(2) Fuse - Potassium chlorate saturated cotton string.
(3) Primary Explosive - Diazodinitrophenol (DDNP), lead picrate, or tetramminecopper (II) chlorate (TAC).
(4) Booster Explosive - RDX and picric acid.

As a direct follow-up to the primary explosives studies conducted during this program, it is recommended that several promising materials be more extensively evaluated in a subsequent program in an effort to obtain a wider selection of primary explosives which can be prepared from a variety of indigenous materials. Specific materials recommended for future study under such a program are:

(1) Lead Azide - A refined field expedient method of preparation should be developed.
(2) Synthetic Applications of Hydrazine Sulfate - The relative ease of preparing this compound and the broad spectrum of preparative possibilities strongly recommend its future investigation.
(3) Polyol Complexes of Lead Chlorate (e.g., glycerol)- The lack of success in this class was mainly due to the failure of initial attempts to prepare basic or normal lead chlorate. It is considered feasible to solve this problem in the future.
(4) Mercarbide and Silver Chlorate.

In addition, it is recommended that future development efforts be directed toward several promising ordnance applications resulting from this program which involve components other than the primary explosive. These include: Booster explosives, linear fusing, contact igniters and certain flare compositions.

The variety of the above described material systems indicates the advantage of permitting some latitude in a development project. Not only is solution of the program goals realized, but other heretofore unknown possibilities become apparent. It is therefore suggested that future investigations be continued in the broader scope of chemistry. Once a feasible candidate is discovered, methods for its preparation can usually be developed after brief, concentrated study.

Appendix

Cartridge Brass Detonator Fabrication

Detonators (blasting cap) can be made from any used small arms cartridge case, .22 cal. or larger. Detonators are used to initiate high explosives (C-4, TNT).

MATERIAL REQUIRED:

Firecracker fuse or string soaked in potassium chlorate
Cartridge case
Primary explosive
Booster explosive
Round wooden stick (small enough just to fit in the neck of the cartridge case)
Knife or drill
Nail
Improvised scale

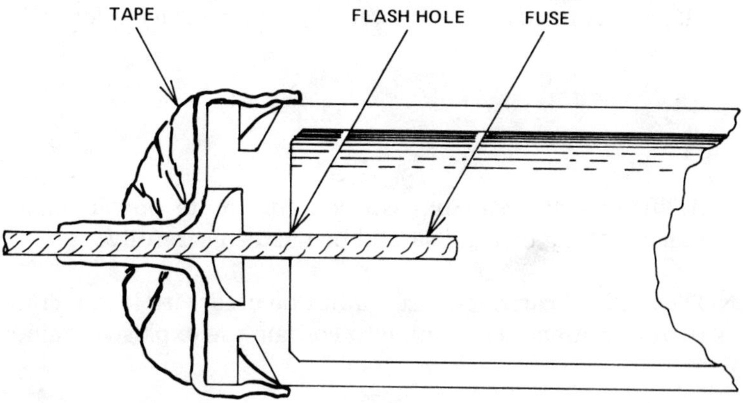

PROCEDURE:

1. Remove fired primer from cartridge case using a sharpened nail.
2. Open out flash hole using a knife or drill; make it large enough to receive fuse.

3. Place fuse in flash hole and tape end of case so fuse will not fall out (see illustration on preceding page).
4. Load appropriate amount of the primary in case:

Primary	Minimum Weight
Lead Picrate	3 grams
Tetramminecopper Chlorate	1 gram
Diazodinitrophenol (DDNP)	0.5 gram

5. Gently tamp primary into case with the wooden stick.
6. Add one gram of the booster. Booster can be RDX (extracted from composition C-4), PETN (from Primacord) or picric acid.
7. Carefully tamp booster into case with wooden stick.
8. If case is not full, fill the remainder with the explosive to be detonated.

CAUTION: Detonator has considerably more power than a military blasting cap and should be handled carefully.

NOTE: .22 cal cartridge case cannot be used with lead picrate as there is not enough volume to contain the explosive train.

ASPATORE

ASPATORE

www.Aspatore.com

Aspatore Books, a Thomson Reuters business, exclusively publishes C-Level executives (CEO, CFO, CTO, CMO, Partner) from the world's most respected companies and law firms. C-Level Business Intelligence™, as conceptualized and developed by Aspatore Books, provides professionals of all levels with proven business intelligence from industry insiders—direct and unfiltered insight from those who know it best—as opposed to third-party accounts offered by unknown authors and analysts. Aspatore Books is committed to publishing an innovative line of business and legal books, those which lay forth principles and offer insights that when employed, can have a direct financial impact on the reader's business objectives. In essence, Aspatore publishes critical tools for all business professionals.

Inside the Minds

The *Inside the Minds* series provides readers of all levels with proven legal and business intelligence from C-Level executives and lawyers (CEO, CFO, CTO, CMO, Partner) from the world's most respected companies and law firms. Each chapter is comparable to a white paper or essay and is a future-oriented look at where an industry, profession, or topic is heading and the most important issues for future success. Each author has been selected based upon their experience and C-Level standing within the professional community. *Inside the Minds* was conceived in order to give readers actual insights into the leading minds of top lawyers and business executives worldwide, presenting an unprecedented look at various industries and professions.

Praise for *Inside the Minds*

"This series provides a practical and focused discussion of the leading issues in law today." – John V. Biernacki, Partner, Jones Day

"*Inside the Minds* draws from the collective experience of the best professionals. The books are informative from an academic, and, more importantly, practical perspective. I highly recommend them." – Keith M. Aurzada, Partner, Bryan Cave LLP

"Aspatore's *Inside the Minds* series provides practical, cutting edge advice from those with insight into the real world challenges that confront businesses in the global economy." – Michael Bednarek, Partner, Shearman & Sterling LLP

"What to read when you want to be in the know—topical, current, practical, and useful information on areas of the law that everyone is talking about." – Erika L. Morabito, Partner, Patton Boggs LLP

"Some of the best insight around from sources in the know" – Donald R. Kirk, Shareholder, Fowler White Boggs PA

"The *Inside the Minds* series provides a unique window into the strategic thinking of key players in business and law." – John M. Sylvester, Partner, K&L Gates LLP

"Comprehensive analysis and strategies you won't find anywhere else." – Stephen C. Stapleton, Of Counsel, Dykema Gossett PLLC

"The *Inside the Minds* series is a real hands-on, practical resource for cutting edge issues." – Trey Monsour, Partner, Haynes and Boone LLP

"A tremendous resource, amalgamating commentary from leading professionals that is presented in a concise, easy to read format." – Alan H. Aronson, Shareholder, Akerman Senterfitt

"Unique and invaluable opportunity to gain insight into the minds of experienced professionals." – Jura C. Zibas, Partner, Lewis Brisbois Bisgaard & Smith LLP

"A refreshing collection of strategic insights, not dreary commonplaces, from some of the best of the profession." – Roger J. Magnuson, Partner, Dorsey & Whitney LLP

"Provides valuable insights by experienced practitioners into practical and theoretical developments in today's ever-changing legal world." – Elizabeth Gray, Partner, Willkie, Farr & Gallagher LLP

"This series provides invaluable insight into the practical experiences of lawyers in the trenches." – Thomas H. Christopher, Partner, Kilpatrick Stockton LLP